On the Go

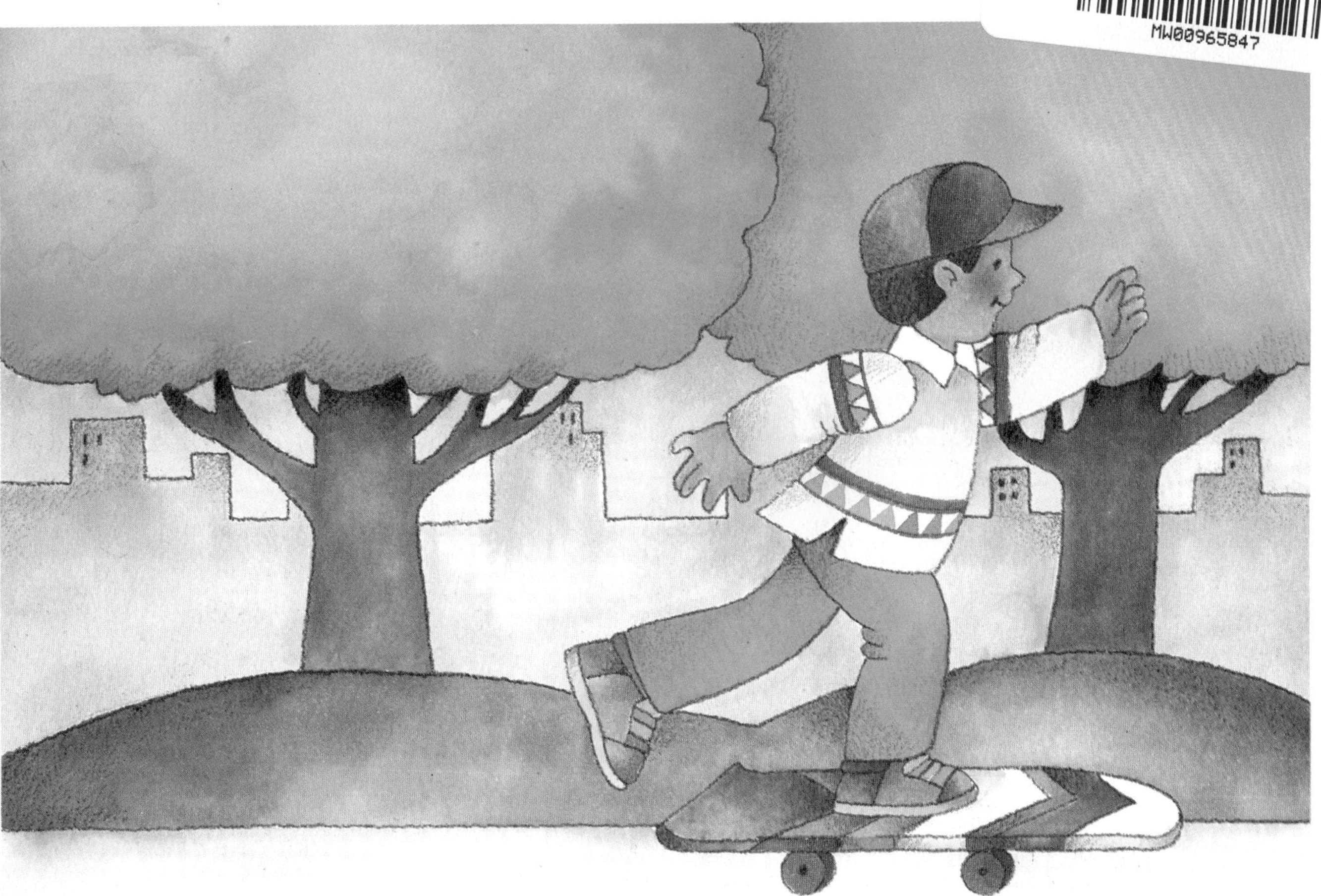

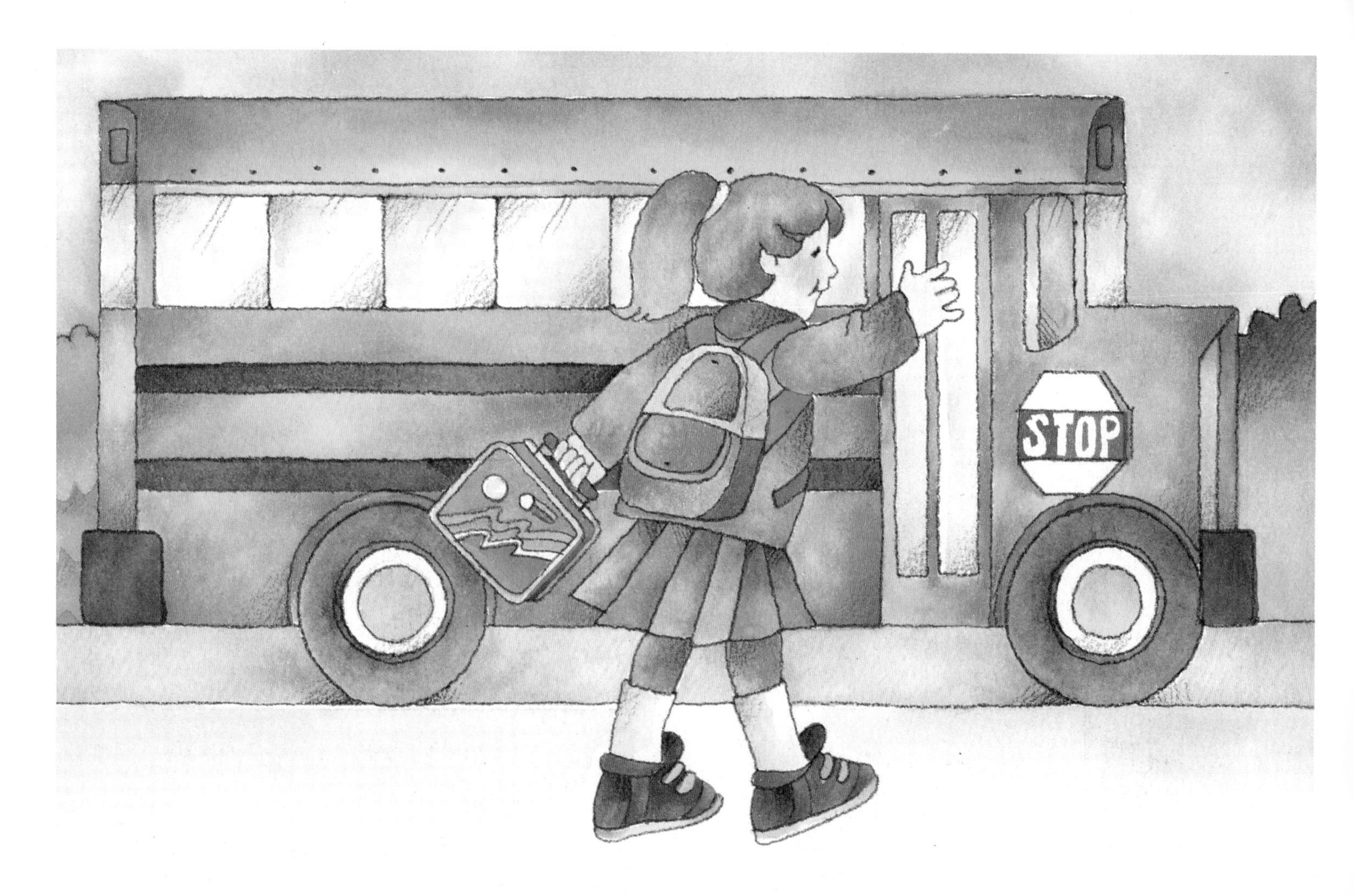

A bus takes me to school.

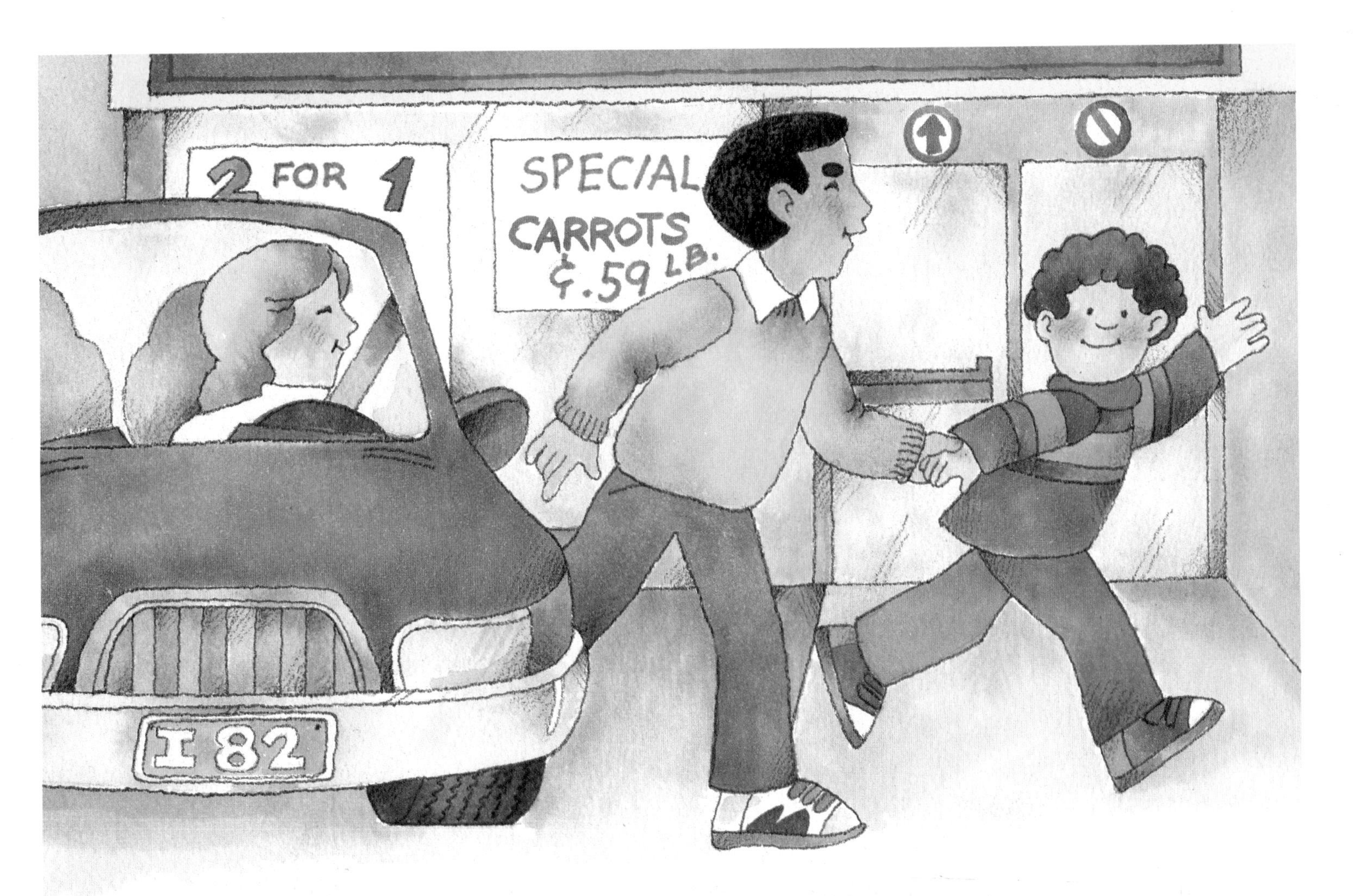

A car takes me to the store.

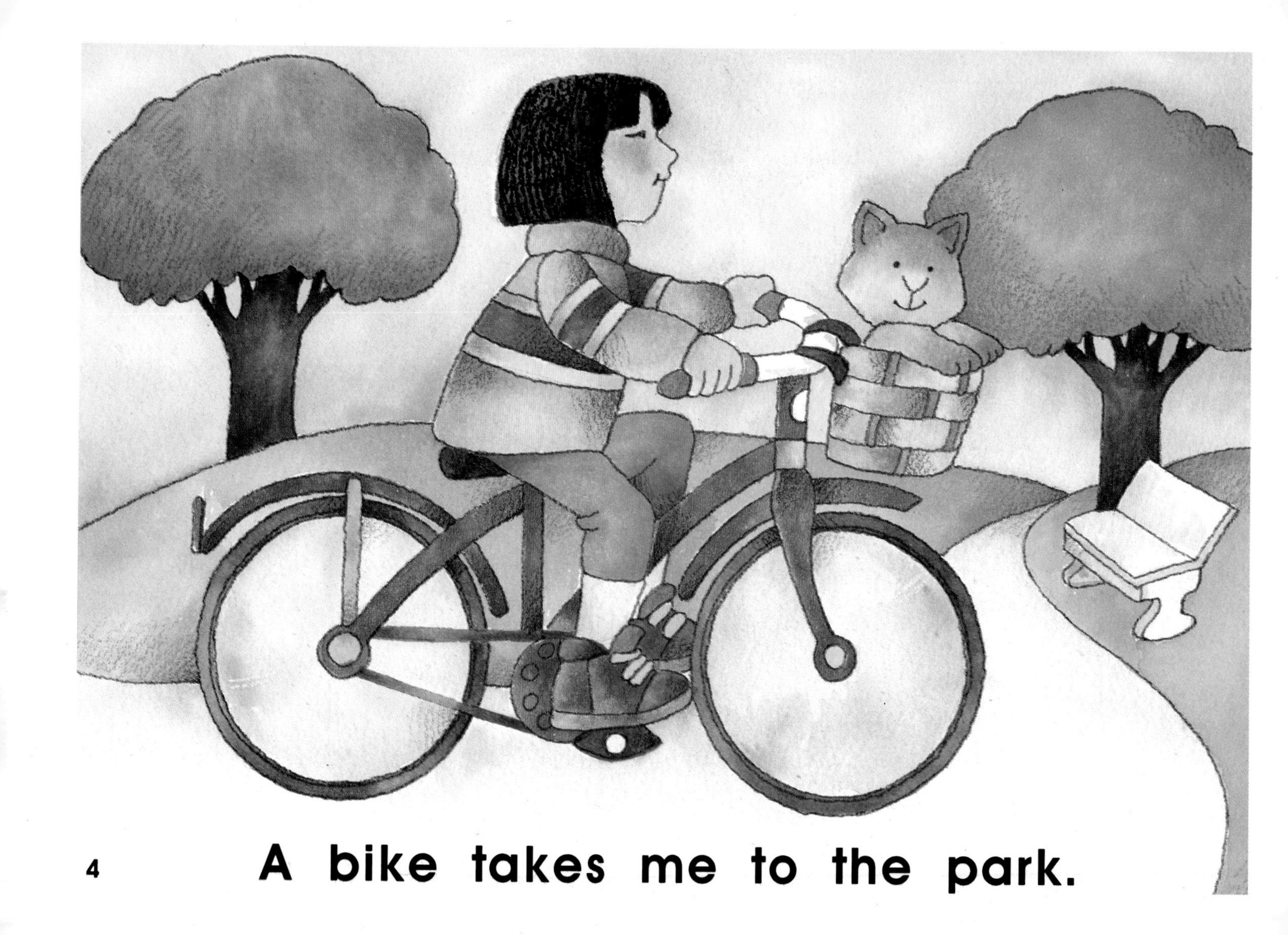

A bike takes me to the park.

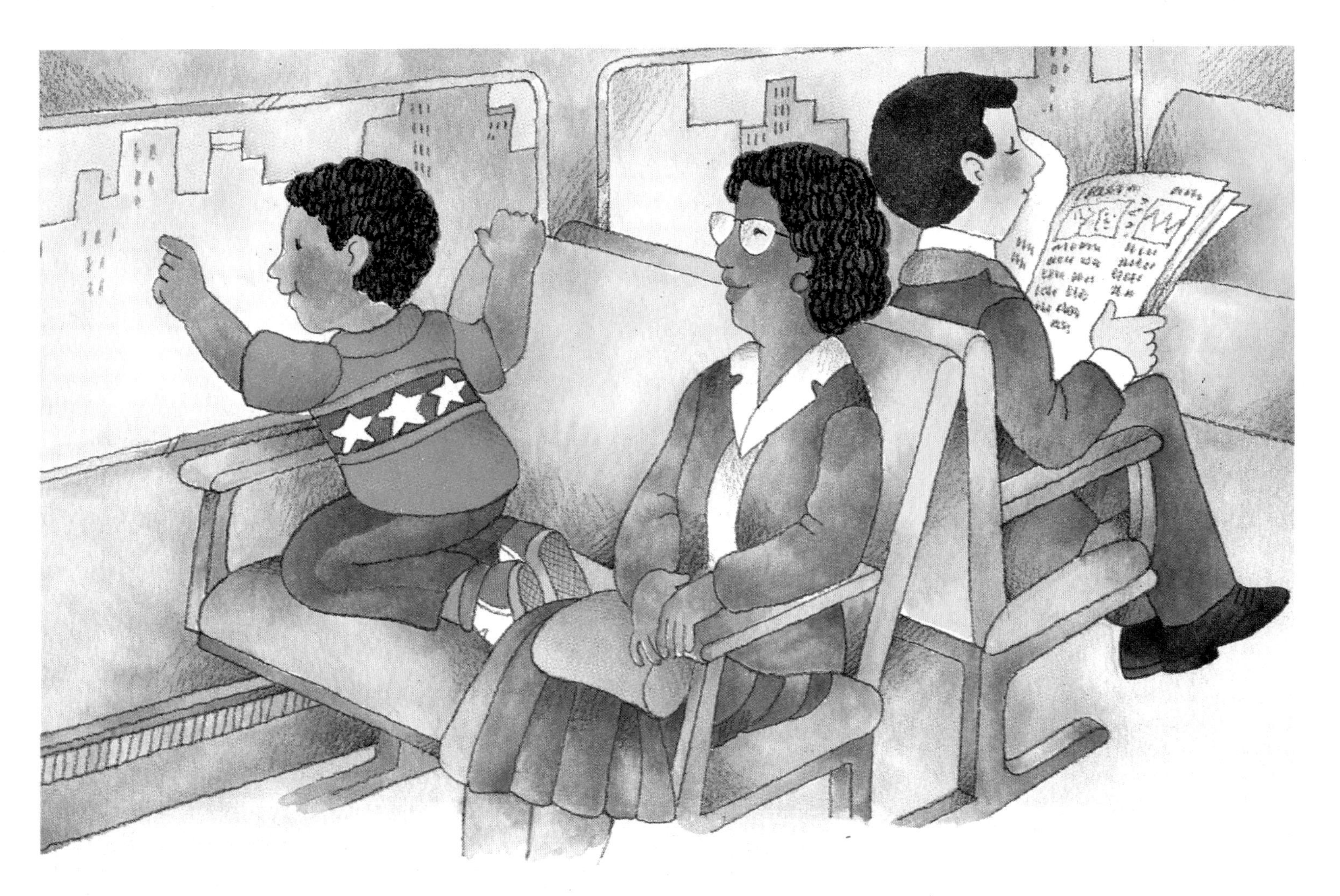

A train takes me to the city.

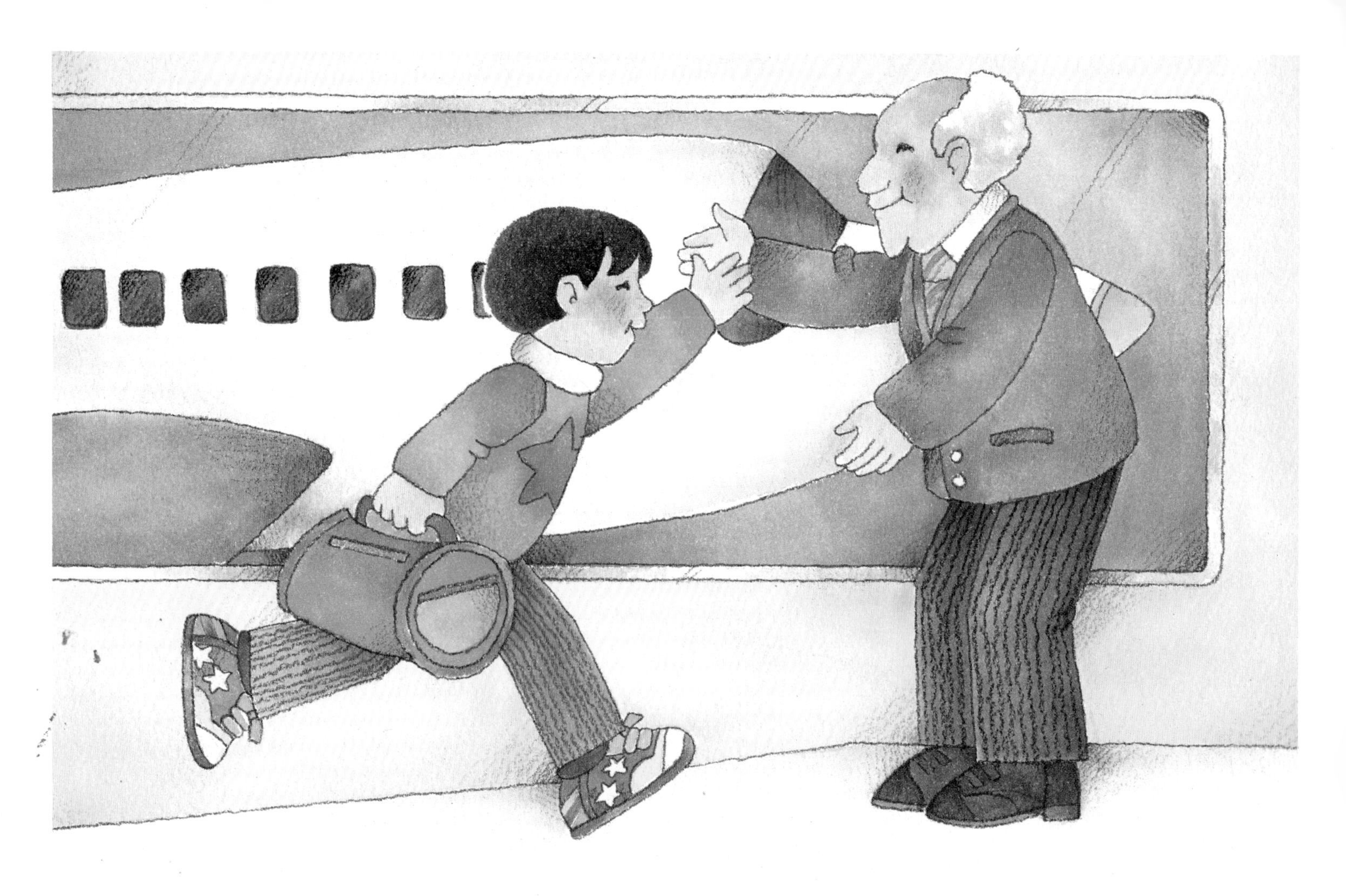

A plane takes me to Grandpa's.

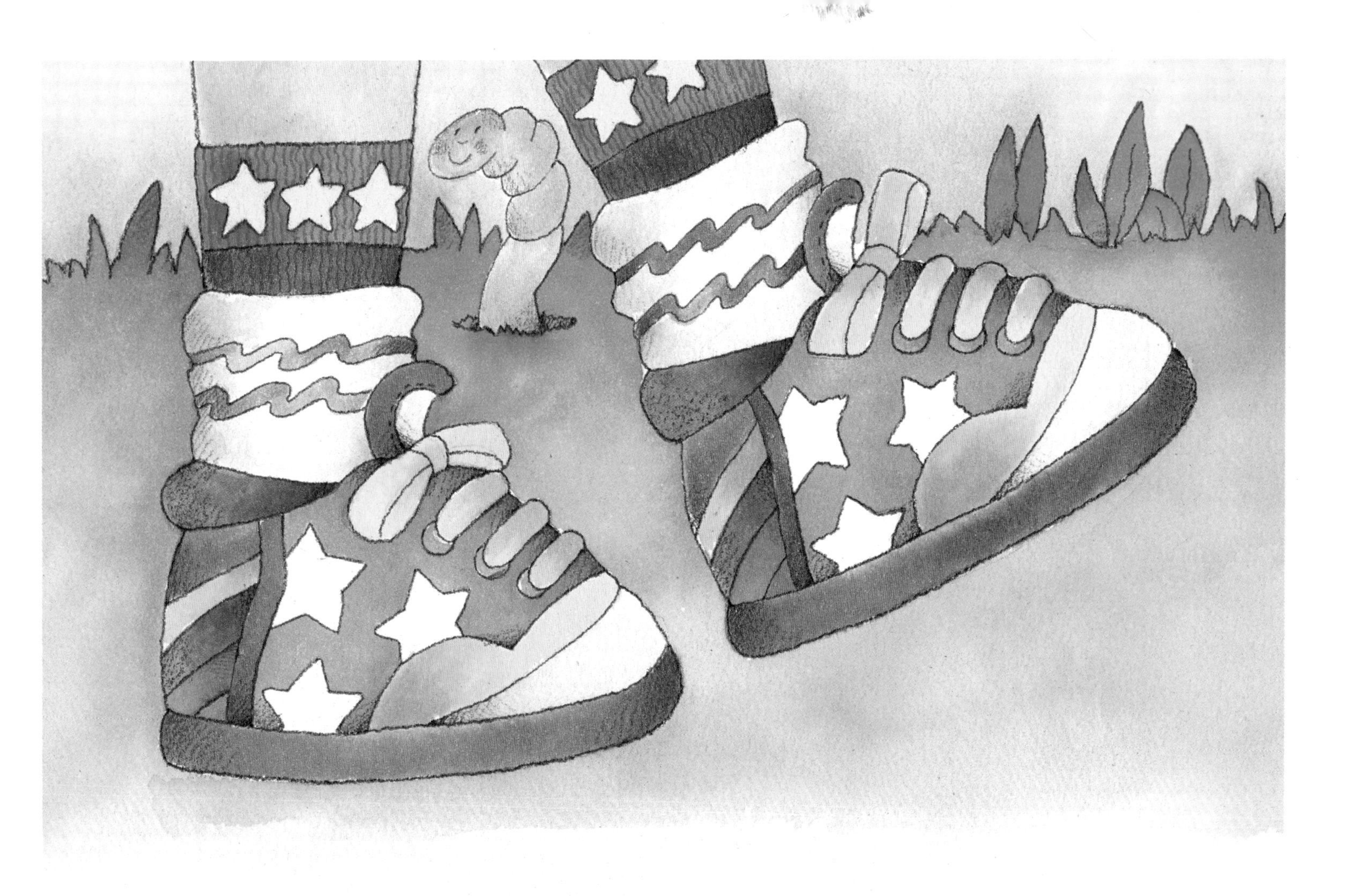

But my feet take me . . .

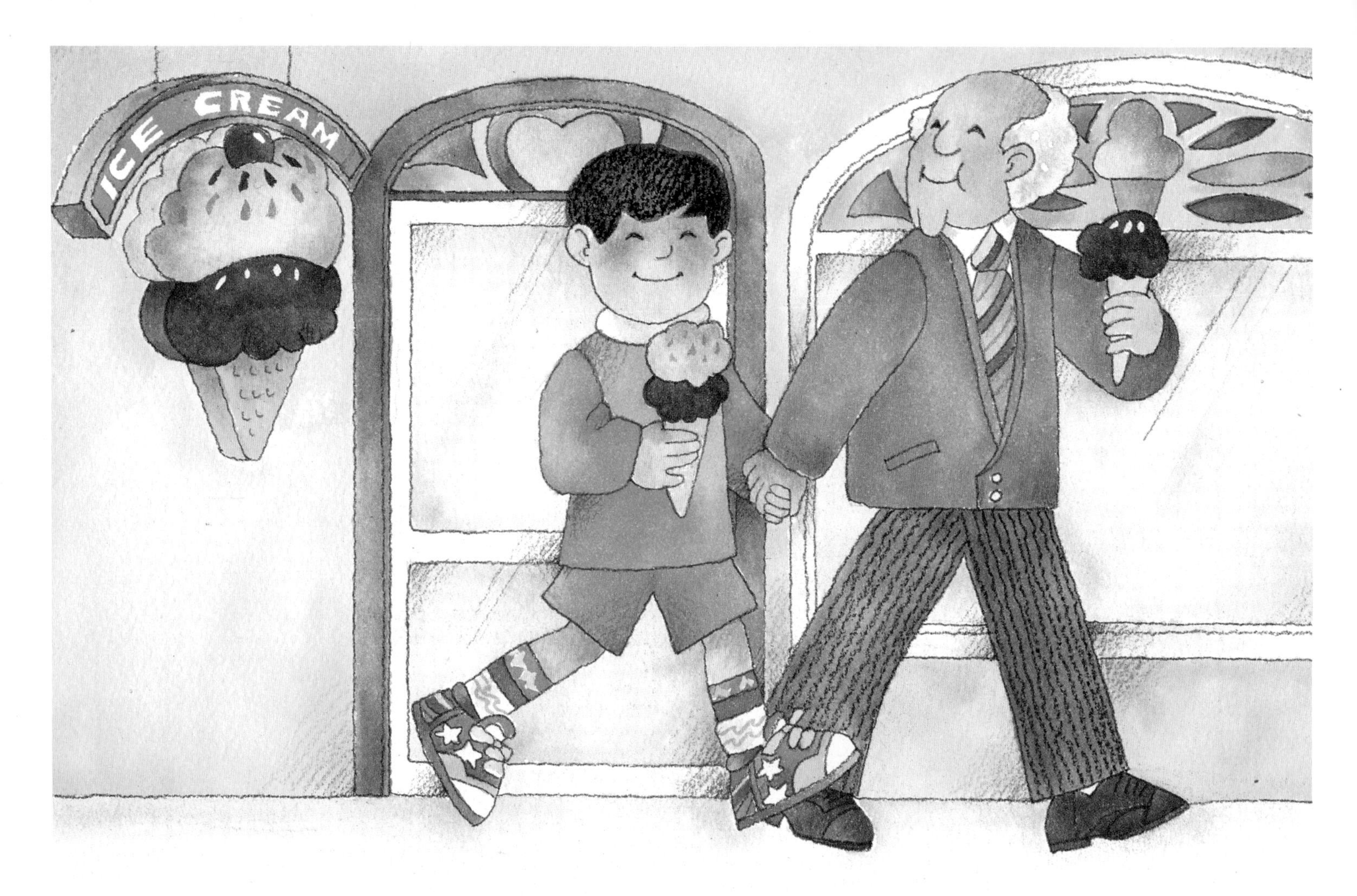

to the ice cream store!